T0173705

THE

LITTLE BOOK OF

Bees

THE

LITTLE BOOK OF

Bees

An illustrated guide to the
extraordinary lives of bees

HILARY KEARNEY

ILLUSTRATED BY AMY HOLLIDAY

HarperCollins*Publishers*

HarperCollins*Publishers*
1 London Bridge Street
London SE1 9GF
www.harpercollins.co.uk

HarperCollins*Publishers*
Macken House,
39/40 Mayor Street Upper,
Dublin 1, D01 C9W8, Ireland

First published by HarperCollins*Publishers* in 2019

5 7 9 10 8 6

Text by Hilary Kearney
Cover and interior illustrations by Amy Holliday
Design by Jacqui Caulton

Hexagon pattern used on cover and pages 2, 6, 8–9, 88–9, 134–5,
156–7, 182–3, 210–11, and 218 © Shutterstock. Label used on
cover and pages 1, 3, 9, 89, 135, 157, 183, and 211 © Shutterstock.
Background used on pages 91 and 124 © Shutterstock.

A catalogue record for this book is
available from the British Library

ISBN: 978-0-00-832427-8

Printed and bound in Latvia

This book is dedicated to anyone who has rescued a bee from their pond, given sugar water to a tired bee, planted flowers in their garden, or told a friend not to swat. Thanks for caring; I hope this book inspires you to do even more for our bee friends.

CONTENTS

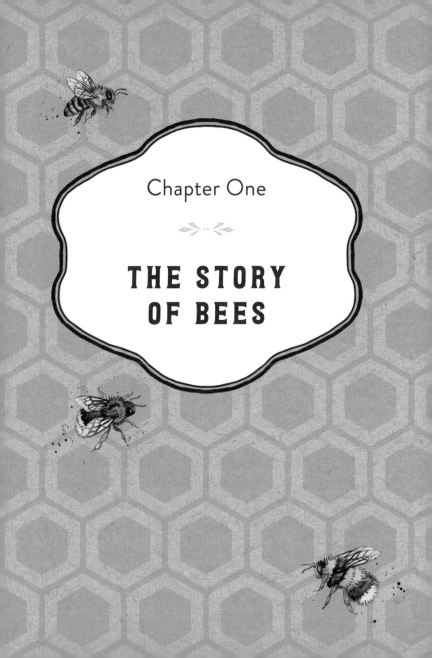

Chapter One

THE STORY OF BEES

FROM DINOSAURS TO DAHLIAS: THE EVOLUTION OF THE BEE

You might be surprised to know that the first bee buzzed alongside a lumbering Stegosaurus. It collected pollen from small, dull flowers without any petals and it most closely resembled a wasp. That's where the story of bees begins, 130 million years ago, in the last epoch of the dinosaurs. Before flowers painted the landscape, before humans walked the Earth.

THE FIRST FLOWERS

It's difficult to imagine a world without flowering plants. They dominate our habitat today, making up 80 per cent of all the plants on Earth. Yet the first flowers emerged relatively recently in geologic time, about 130 million years ago, around the arrival of the first bees.

When early flowers appeared among the ferns and pine trees of the Cretaceous period, they looked nothing like our flowers today. There were no showy dahlias standing colourful and bright or stalks of shapely foxgloves made heavy with nectar. The first flowers were modest, without ornament, and hardly distinguishable from the plants around them. Until this point, most plants reproduced by scattering their pollen to the wind, but nature was about to find a more efficient method of pollination: insects.

Hungry beetles were among the first to seek out the nutrient-rich pollen that flowers provided. The pollen clung to their wiry hairs and came with them as they travelled from flower to flower. This method of transferring pollen was more direct and less wasteful than wind pollination. Its early success led to the rapid expansion of flowering plants and the insects that pollinated them.

EXPLOSIVE CHANGE

Flowers began to spring up everywhere and quickly developed insect-attracting features to distinguish them from their neighbours. Water lilies and magnolias were the first to unfold generous, white petals. The eye-catching petals made it easier for pollinators to find them and also served as convenient landing pads. Other flowers wooed insects by developing alluring scents, bright colours, and finally the sweet reward of nectar.

As flowers became more complex and diverse, so did the pollinators that visited them. They evolved not just alongside one another, but in response to each other. Flowers strived to be more inviting to insects and insects became better and better pollinators. Some pollinators grew long, tubular tongues for extracting nectar, others developed a keen sense of smell or a more acute perception of colour. And for every elongated tongue, there was a corresponding deep-throated flower. Pollinators evolved to suit specific flowers and became specialists.

THE BEE ARRIVES

Somewhere in this explosion of life, the bee buzzed into existence. This primordial bee was actually a predatory wasp, who would change everything by switching to a vegetarian diet. Imagine a wasp stalking its prey from among the newly minted pollinators of the time. The wasp likely ate a pollen-dusted beetle for lunch and began to develop a taste for the pollen rather than the beetle. Eventually, the wasp gave up its predatory ways and pollen became its only source of protein.

The first bees further distinguished themselves from wasps by developing various pollen-collecting methods. They sprouted feathery hairs on their legs and abdomens that could be packed with pollen and flown back to the nest where it would be used to feed their offspring. When nectar became available, bees found the sugary substance to be an excellent source of energy. They grew long tongues and sipped on it while gathering pollen.

BEE FACT

ALTHOUGH IT'S COMMONLY BELIEVED THAT BEES EVOLVED FROM POLLINATING WASPS, THE DETAILS OF THE THEORY ARE STILL SHROUDED IN MYSTERY BECAUSE OF GAPS IN THE BEE FOSSIL RECORD. THE MOST RECENT CLUE WAS UNCOVERED IN 2006, WHEN THE OLDEST KNOWN BEE, *MELITTOSPHEX BURMENSIS*, WAS FOUND ENCASED IN AMBER. THIS 100-MILLION-YEAR-OLD FOSSIL PROVIDES NEW EVIDENCE IN SUPPORT OF THE THEORY BECAUSE IT HAS TRAITS OF BOTH

BEES AND WASPS. IT'S CONSIDERED
PARTICULARLY IMPORTANT
BECAUSE IT ALSO HAS BRANCHED
HAIRS, WHICH TODAY'S BEES USE
FOR COLLECTING POLLEN.

Over time, some bees became social creatures and gave up their solitary lifestyle in favour of colony life. They developed complex societies where many bees shared a single nest and the duties of maintaining it. They evolved special stomachs for carrying nectar and built structures in their nests to store it. Eventually, this practice led to the creation of honey.

Bees emerged as the most successful pollinators of the era and still hold the title today, with more than 20,000 species around the globe.

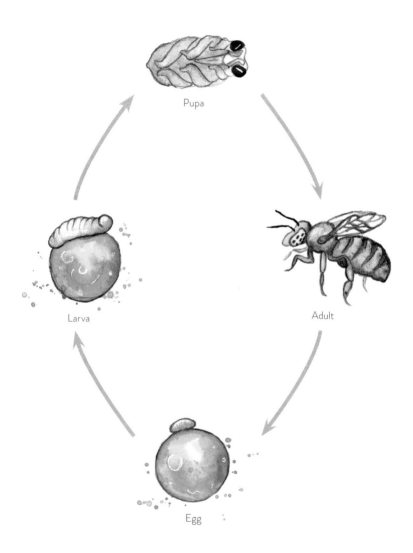

Pupa

Larva

Adult

Egg

FROM EGG TO BEE: THE BEE LIFE CYCLE

Every bee starts as an egg, but it doesn't crack open like a chicken egg when it's time to hatch. Instead, its soft membrane dissolves and its nutrients are absorbed by a tiny larva. In the larval stage, the helpless grub must eat a tremendous amount of food. Mother bees are charged with collecting a supply of pollen and nectar so that their young have the energy to grow. A bee larva will shed its skin five times before entering its final period of development: the pupal stage. However, some bee species spend months or even years in a larval state before going through metamorphosis. Just like the caterpillar of a moth, the bee larva spins a cocoon before transforming itself. Inside the cocoon, the larva develops a head, legs, and body before finally sprouting its hair and wings. Once the change is complete, the young bee chews its way out of the cocoon and commences its adult life.

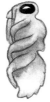

Among solitary bees, the adult's primary duty is to mate. Females seek out nesting sites and begin to store provisions for their future young. As the females busy themselves with the task of collecting pollen and nectar, expectant males jockey for a chance to mate with them among the most popular flowering plants. Clusters of male bees can sometimes be found sleeping overnight inside flowers during mating season. Back at her nest, the female bee amasses pollen until she has enough to feed her young through its larval state. She often constructs a pillow-like pollen patty and lays her egg on top of it. Most bee species live only one year. They emerge in spring, raise their offspring through the summer, and then die. Some will lay just one egg in a season, while others produce many in chambered nests or separate tubular cavities.

In contrast, social bees have evolved to share nesting duties. They live in colonies and not all members are capable of mating. Instead, their reproductive duties are delegated to the males and a singular queen bee. The other colony members are called workers and are tasked with all the non-reproductive activities necessary for survival, such as foraging for food.

POLLINATION

Pollen is nature's glitter: it gets on everything. This is by design, of course. Flowers want their pollen to be spread about because it's how they reproduce. When a flower's male part

(the stamen) produces pollen, it must be transferred to the flower's female part (the pistil) for the plant to make seeds. The pollen can't travel on its own, however, it needs the help of animals, wind, or water.

BEE FACT

DID YOU KNOW THAT STATIC
ELECTRICITY PLAYS A ROLE IN
POLLINATION? AS SHE FLIES FROM
FLOWER TO FLOWER, A BEE'S BODY
GENERATES A POSITIVE ELECTRICAL
CHARGE FROM THE FRICTION OF
MOVING THROUGH THE AIR. FLOWER
PETALS, IN CONTRAST, USUALLY GIVE
OFF A NEGATIVE CHARGE. THE RESULT
BEING THAT WHEN A POSITIVELY
CHARGED BEE APPROACHES A
NEGATIVELY CHARGED FLOWER, THE
POLLEN LEAPS FROM THE FLOWER TO
THE BEE. SOMETIMES THIS HAPPENS
BEFORE THE BEE HAS EVEN LANDED.

BEES IN YOUR GARDEN

In the garden, bees and other pollinators, such as wasps, flies, butterflies, moths, and birds, are helpful friends. This is especially true if you are growing fruits and vegetables. When your edible plants are in bloom, each flower has the potential to become a fruit or vegetable, but it must be pollinated. Gardens with a healthy bee population will not only have a greater yield, but the more a flower is pollinated, the bigger and better-tasting the fruit will be.

BEES IN AGRICULTURE

In agriculture, bees are prized pollinators. Some crops, including blueberries, cherries, and almonds, are almost entirely dependent on bees for pollination. When these crops are in bloom, farmers can't risk a low pollination rate by relying only on their local bee population. For this reason, many farmers contract with beekeepers to bring honey bee colonies to their land. Some farmers also purchase bumble bee colonies to pollinate inside greenhouses where honey bees will not go. Without the help of these bees, the farmers may not yield enough food to stay in business. In North America, migratory beekeepers will travel nearly year-round, crossing from one side of the USA to the other with hundreds of honey bee colonies.

Unfortunately, migratory beekeeping has some drawbacks. Critics say the large influx of honey bees can negatively impact native bee populations. Honey bee advocates also worry that this practice spreads honey bee pests and diseases to regions previously unaffected.

BEE FACT

FLOWERS USE ALL THEIR WILES TO ATTRACT BEES, AND THE BRIGHTLY COLOURED PETALS AND INTOXICATING SCENTS ALSO WORK THEIR MAGIC ON HUMANS. BUT SCIENTISTS HAVE DISCOVERED THAT SOME FLOWERS EMPLOY ANOTHER METHOD, UNSEEN BY THE HUMAN EYE. THESE CLEVER FLOWERS DISPLAY STARK PATTERNS VISIBLE ONLY IN THE ULTRAVIOLET LIGHT SPECTRUM TO GUIDE BEES TO THEM.

A flower as it appears
in the ultraviolet light
spectrum visible to bees

A flower as it appears
to the human eye

POLLINATOR DIVERSITY

Even though honey bees are the most popular pollinators in agriculture, it's important to support a diversity of bee pollinators. In some cases, one type of bee may not be capable of pollinating a particular species of plant. Honey bees, for example, are unable to pollinate tomatoes. Tomato flowers need to be vibrated strongly in order to shake loose their tightly packed pollen. Honey bees cannot 'buzz pollinate', but bumble bees and other large species excel at this method. Honey bees have also been shown to pollinate more efficiently when working alongside other bee species. The presence of wild bees disrupts the honey bee foraging routine and causes them to visit more flowers than they normally would. We can help support a diverse group of bees by growing native flowering plants alongside our favourite foods.

ALTHOUGH SOME PLANTS SELF-POLLINATE, MOST ENGAGE IN CROSS-POLLINATION AND RELY ON AN OUTSIDE AGENT TO MOVE THEIR POLLEN. THESE AGENTS CAN BE DIVIDED INTO TWO CATEGORIES: LIVING AND NONLIVING. THE VAST MAJORITY OF LIVING POLLINATORS ARE INSECTS, BUT BIRDS AND BATS ALSO SERVE AS MAJOR POLLINATION VECTORS.

NONLIVING AGENTS INCLUDE WATER
AND WIND. ABOUT 12 PER CENT
OF THE WORLD'S FLOWERING PLANTS
ARE WIND POLLINATED. OTHER
PLANTS USE WATER TO FLOAT
THEIR POLLEN DOWNSTREAM
TO WAITING FLOWERS.

THE BEE FAMILY TREE

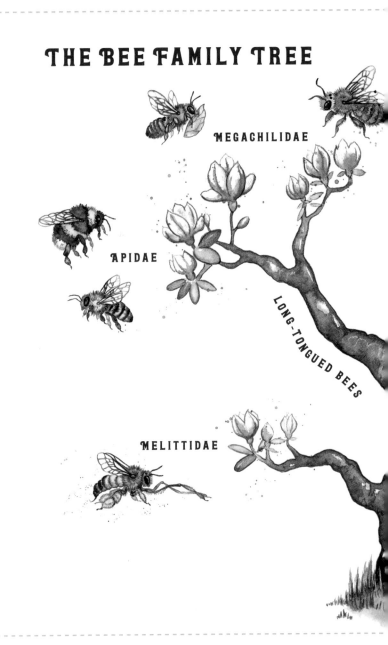

MEGACHILIDAE

APIDAE

LONG-TONGUED BEES

MELITTIDAE

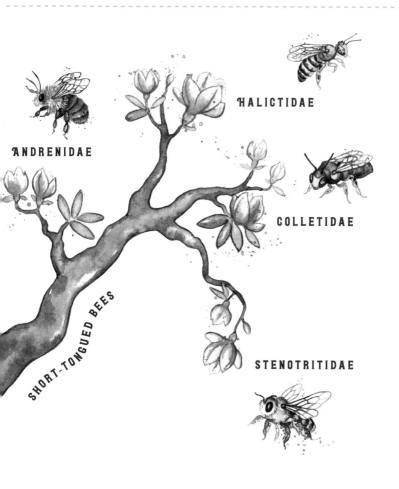

HALICTIDAE

ANDRENIDAE

COLLETIDAE

SHORT-TONGUED BEES

STENOTRITIDAE

THE BEE FAMILY TREE

For most people, the word 'bee' conjures an image of a honey bee or bumble bee. Yet there are 20,000 different kinds of bee in the world – more species than of mammals and birds combined – categorized into seven different families by some shared traits.

SPECIES SPOTLIGHT

The majority of bee species are classified as solitary bees. They don't have colonies or queen bees and they don't make honey. They find a mate, build a small nest, raise their young, and then die. Although they are largely unrecognized, solitary bees are important pollinators. They work alongside the more popular species to pollinate native plants, crops, and home gardens. Within each bee family section in the following pages, a particular species has been highlighted in the 'Species Spotlight' as an example of the family and its relevant traits and varieties.

ANDRENIDAE FAMILY

The largest bee family, Andrenidae, or mining bees, is named for its nesting behaviour – they dig tunnels in the soil. Although they are solitary bees, some species of mining bee are gregarious and like to nest close to one another. They form little bee neighbourhoods with many tunnels dug side by side.

Species Spotlight

ASHY MINING BEE
(Andrenidae family)

The ashy mining bee is well known in the UK. In early spring, females dig their nests in bare garden borders, lawns, paths, and other open spaces. They can be recognized by the little piles of dirt that accumulate next to their nest holes. They are extremely docile and healthy for lawns because their tunnels aerate the soil. Their striking colouring of grey and black makes them easy to recognize; the males in this species have a dense collection of grey hairs on their face, giving them the moustached appearance of a proper English gentleman (see page 41)!

APIDAE FAMILY

The agriculturally significant Apidae family is distinguished by many standout species. The group includes honey bees, bumble bees, and carpenter bees, as well as orchid bees and squash bees, among others. Bees in the Apidae family tend to be stout-bodied with a dense covering of hair.

Species Spotlight
BLUE-BANDED BEE
(Apidae family)

An Australian native, the blue-banded bee is easily recognized by its dazzling turquoise stripes, but you may have trouble getting a good look – they are extremely fast fliers! These bees are responsible for pollinating a significant amount of Australia's crops as well as many native plants using the buzz-pollination method often credited only to non-native bumble bees. Females nest close to one another in clay, soft rock, or sometimes in the mortar between bricks. Males, too, can be found in groups: they roost together overnight, lining up in a row as they cling to a single plant stem. Curiously, the blue-banded bee seems to prefer blue flowers and is sometimes attracted to people wearing the colour blue.

Species Spotlight

CARPENTER BEE
(Apidae family)

Carpenter bees are gentle giants, named for their habit of nesting in decaying wood. Most are dark in colour with little hair, but many have a gorgeous blue sheen to them, and one Asian species (*Xylocopa caerulea*) is covered in a striking cerulean fuzz. Unlike other bees, who live for only one season, carpenter bees can live for several years. Female carpenter bees like to nest close together, chewing shallow tunnels in which to raise their young, and males often patrol these nest sites in an attempt to scare off anyone who comes near; but don't worry – males don't have stingers! Although carpenter bees are scorned by humans as pests, they are pollinators of some important crops, including cotton, tomato, aubergine, and passionfruit.

Xylocopa caerulea

Species Spotlight

ORCHID BEE
(Apidae family)

Orchid bees live in the tropical regions of Central and South America. These strikingly beautiful metallic bees come in a rainbow of colours, resembling gemstones more than bees. Although male bees don't usually participate in pollination, some species of orchid are pollinated exclusively by male orchid bees. That's because male orchid bees have a curious habit of concocting their own cologne. They visit orchid flowers, but also fungi, and even rotting vegetation, to collect scents that they store on their hind legs. It's not clear why they collect scent compounds, but it may play a role in attracting a mate.

Species Spotlight

SQUASH BEE
(Apidae family)

Found from North to South America, the squash bee limits its foraging to pumpkins, gourds, and squash. Although they closely resemble honey bees, squash bees are slightly larger and possess shaggy leg hair designed specifically to collect the coarse pollen grains of squash flowers. Unlike other bees who thrive in warm daylight, squash bees are most active just before dawn, zipping between blossoms in the twilight. In the afternoon hours, you can find them even more easily – males nap the day away at the bottom of the blossoms.

COLLETIDAE FAMILY

Bees in the Colletidae family are often called plasterer bees because they coat their underground nesting cavities with a cellophane-like lining. Many of the bees in this family do not carry pollen on their legs, but instead use a stomach-like organ that enables them to carry the pollen inside their bodies.

Species Spotlight
MASKED BEE
(Colletidae family)

Masked bees are a segment of plasterer bees, named for their distinctive white or yellow facial markings. They have slender, sparsely haired bodies and tend to be quite small. They can be found all over the world, but are notable for their success as island dwellers. They are the only bees native to Hawaii, USA, which used to boast 60 different species of masked bee. Sadly, because of habitat destruction, many of these are now extinct or under threat. There are currently seven types of Hawaiian masked bee on the US Endangered Species List.

HALICTIDAE FAMILY

Halictidae are commonly called sweat bees, as they are attracted to the salt in human perspiration. Like honey bees, some species of sweat bee are eusocial and live in organized colonies that share the labour of raising their young and rely on a queen bee for reproduction.

Species Spotlight
METALLIC GREEN BEE
(Halictidae family)

Metallic green bees are common inhabitants of gardens in the USA. They come in many shades of metallic blues and greens that dazzle in the sunlight. Although they forage on a variety of plants, they prefer flowers in the sunflower family. Males are easily recognized by their distinctive yellow-striped abdomens, while the females are entirely green. Most are solitary and nest in the ground. They are attracted to gardens with bare patches of soil, free of mulch, for nesting.

BEE FACT

THE SMALLEST BEE IN THE WORLD (*PERDITA MINIMA*) MEASURES JUST 2 MM (¹⁄₁₆ IN) – THAT'S SMALLER THAN MOST ANTS – WHILE THE WORLD'S LARGEST BEE (*MEGACHILE PLUTO*) IS ALMOST 20 TIMES BIGGER AND BOASTS A WINGSPAN OF 63.5 MM (2½ IN). ALSO KNOWN AS THE 'FLYING BULLDOG', IT IS SO RARE THAT IT WAS ONCE THOUGHT EXTINCT, WITH NO DOCUMENTED SIGHTINGS FOR 38 YEARS. IT HAS SINCE BEEN REDISCOVERED ON SEVERAL INDONESIAN ISLANDS, BUT REMAINS ELUSIVE.

Honey bee

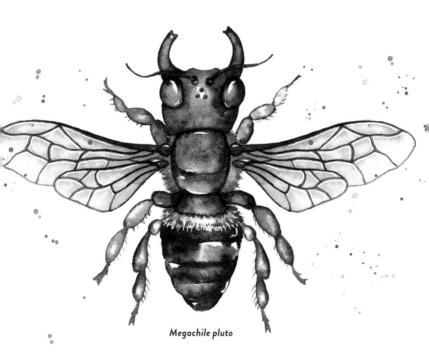

Megachile pluto

Perdita minima is a fraction of the size of a honey bee, seen here
for comparison against the *Megachile pluto*.

MEGACHILIDAE FAMILY

Bees in the widespread Megachilidae family – mason bees, leafcutter bees, and carder bees – are known for their ingenious nests. The family is often divided into segments with their common names reflecting these fascinating nesting habits. The bees in this family do not carry pollen on their legs. Instead, females have a dense collection of hairs on the underside of their abdomens where they pack on pollen.

Species Spotlight

LEAFCUTTER BEE
(Megachilidae family)

Leafcutter bees nest in the same way as mason bees, but instead of working with mud, they build with carefully cut pieces of leaf. A leaf-lined cavity may house several developing bees with leaf partitions between. Female leafcutter bees use their large mandibles to cut and carry pieces of leaves back to the nest. If you've ever noticed neat, circular holes along the edges of the leaves in your garden, you may be hosting some leafcutter bees. Some leafcutters may even take clippings of flower petals. In Turkey, one rare species (*Osmia avosetta*) constructs its cocoon entirely from flower petals, digging shallow tunnels in the ground to house its gorgeous multi-coloured floral sculptures that are used to raise its young.

Osmia avosetta flower cocoon

Species Spotlight

MASON BEE
(Megachilidae family)

Mason bees are named for their habit of using mud to construct their nests. They like to build in existing cavities, such as old insect burrows, dried stems, rock crevices, and even empty snail shells. Mason bee nests are linear: the eggs are laid in a row, separated by mud partitions. Once the female has filled a cavity with her eggs, she seals it with a mud plug and begins her search for a new nesting site. A single female may establish several nests in a season. Mason bees have special horns or corkscrew hairs on their heads that allow them to pollinate hard-to-reach flowers. They are favoured in commercial agriculture and are especially skilled pollinators of apples, cherries, and other fruit trees. For this reason, they are sometimes called orchard bees.

MELITTIDAE FAMILY

The Melittidae family is a small, ancient family of only 200 species found primarily in Africa. The oldest known fossilized bee, which dates back 100 million years, is thought to be from the Melittidae family, suggesting that bees, like humans, may have originated in Africa.

Species Spotlight
OIL BEES
(Melittidae family)

Some South African species in the Melittidae family collect floral oils as well as pollen. The oil is mixed with the pollen and used to feed their larvae. A few of the females who exhibit this behaviour have evolved extra-long forelegs to reach the oil sacs found deep in certain flowers. This adaptation is so extreme, some species have forelegs that are twice the length of their bodies!

STENOTRITIDAE FAMILY

Found only in Australia, the Stenotritidae family is the smallest of the bee families with only 21 species. These large, hairy bees are extremely fast fliers who nest in the ground. They line their underground nesting chambers with a waterproof secretion that has a waxy appearance.

Species Spotlight
GUM BEES
(Stenotritidae family)

Although *Stenotritus pubescens* lacks a common name, we would do well to call them gum bees because of their preference for foraging on Eucalyptus or 'gum' tree flowers. These bees are early risers and are often the first bees to begin foraging in the morning. Flowering gum trees whir with the activity of female gum bees who are quick and noisy fliers. They have been observed to nest in groups, digging their tunnels in the soil only a short distance from gum tree groves.

CUCKOO BEES
(Found in a number of bee families)

If you saw a cuckoo bee, chances are, you'd mistake it for a wasp. These sleek bees have little use for hair, because they don't engage in the hard work of collecting pollen like other bees. They cheat the system. Most mother bees spend weeks collecting pollen to feed their young, but cuckoo bees avoid this work by sneaking into the nests of other bees to lay their eggs. After a cuckoo bee egg hatches, the larva uses a set of deadly mandibles to eliminate its competition. With the original bee's progeny dead, the cuckoo bee has the pollen patty all to itself.

Cuckoo bees are a type of bee found in several bee families. They are categorized by 'kleptoparasitic behaviour' – stealing the nest of other bees. So far, cuckoo bee species have been discovered in the families Apidae, Halictidae, and Megachilidae.

BEE FACT

WITH SO MANY SPECIES OF BEES, EVEN SCIENTISTS CAN HAVE TROUBLE IDENTIFYING THEM ALL. IN 2014, THE US GEOLOGICAL SURVEY COLLECTED AN UNKNOWN SPECIES OF BEE IN ARGENTINA, BUT HAD SO LITTLE INFORMATION ON IT THAT THEY OFFICIALLY LISTED ITS GENUS AND SPECIES AS 'BEE CUTE FURRY FACE'.

BEE ANATOMY

Despite the incredible range of size, colour, and behaviour found among bee species, they share a general anatomy. A bee's body can be divided into three segments: the head, the thorax, and the abdomen. They have six legs, two pairs of wings, a set of antennae, and five eyes.

SIMPLE EYES (OCELLI)

Bees also have three small eyes in a triangula formation on the top of their head. Unlike th compound eyes, these simple eyes have just one lens. They are used mainly to navigate and orient the bee in relation to the sun.

ANTENNAE

A bee's antennae are used primarily for smell. They are covered with microscopic pits and hairs, called sensilla, that help them to detect and process the many fragrances they encounter, especially floral scents.

COMPOUND EYES

The two large eyes on either side of a bee's head are called compound eyes. They are made up of thousands of individual lenses, each one creating a tiny point of light and colour that will make up a picture of the world. The end result is reminiscent of a Pointillist painting, with many small dots making up a larger image. Bees also experience colours differently – they can see hues of greens, blues, and purples, but they cannot see the colour red. They can also see ultraviolet light.

MANDIBLES

A bee's mandibles (or jaws) are useful tools. They are used to fight with other bees, carry nesting materials, and to cling to plant stems.

PROBOSCIS

The proboscis is the bee's tongue. Bees are often divided into two groups: long-tongued bees and short-tongued bees. The proboscis is used to collect nectar and is often described as being similar to a straw that allows the bee to suck up nectar, but this is really only accurate for long-tongued bees. Short-tongued bees cannot create much suction with their tongue; instead, they use it to lap up the nectar, much like a cat.

LEGS

Most bees use their legs to collect and carry pollen, but depending on the species the legs may have very different features. A bee's legs are like a Swiss army knife, with many notches, hooks, and hairs designed for specific tasks. In most species, the front legs have a special notch that's used to clean the antennae, while their back legs have a kind of brush that's used to clean their wings.

FOREWING/HINDWING

At first glance, a bee may appear to have only two wings, but they actually have four. The two sets clip together with tiny hooks, making them appear to be one structure. In flight, the wings of a bee twist and rotate in a figure-of-eight pattern that allows them to hover and carry heavy loads. If you've ever watched a bee take off from a flower petal, you might have noticed that they resemble a tiny helicopter – they seem to lift straight up into the air.

TERGA

A bee's abdomen is covered in a series of protective plates collectively called terga. These telescoping segments protect the bee's organs while still allowing for mobility.

SPIRACLES

Ever wonder how a bee breathes? Bees have special openings along the sides of their bodies called spiracles that allow them to vent air directly to their organs through a branching system. Honey bees can often be seen pumping their abdomens after landing from a long flight; this is their way of catching their breath.

STINGER

The stinger was once a tubular organ used for laying eggs. Over time, it evolved into a defensive tool and is now so modified from its original purpose that eggs no longer pass through it at all. Instead, eggs emerge from a separate opening near the base of the stinger. Since the stinger was initially used for laying eggs, male bees lack it entirely and cannot sting.

SCOPAE

Scopae is the name for a collection of hairs on female bees designed to collect pollen. Most bees have them on their back legs, but some (such as the red mason bee shown here) have them on the underside of their abdomen. The hairs are like a dense brush designed to attract and transport dry pollen.

POLLEN-COLLECTING TOOLS

Female bees have several different methods for collecting pollen. Each species uses a unique combination of physical tools and behaviours to gather and transport this nutritious food. If you stop to observe them, you'll find that some bees roll playfully between the anthers, while others bite and pull at them deliberately. Most bees amass the pollen on their back legs, but some have additional structures that aid them in gathering pollen from specific flowers, including an extra tuft of hair on their head or hooked hairs that grab pollen from deep within the flower. Despite these specialized traits, bees tend to have two main features to help them collect and carry pollen back to their nests.

CORBICULAE

Honey bees, bumble bees, and other species have a special bowl-like indentation on their back legs designed to transport large mounds of pollen. They comb the pollen from their body hair and mix it with saliva and nectar before packing it into the corbiculae or 'pollen basket'. When full, these colourful leg accessories can grow quite large, extending well beyond the parameters of the leg.

SCOPAE

Depending on the species, a bee's scopae can be found either on her back legs or on her belly (as with the red mason bee, shown left). Although the scopae is characterized as a dense gathering of hairs, the exact thickness varies by species. Squash bees, for example, have sparsely placed hairs on their legs that are best for gathering the large pollen grains of squash flowers. In contrast, small sweat bees, like the one above, have finer hairs, capable of trapping pollen from a greater variety of blossoms.

HOME SWEET HOME: BEE NESTING BEHAVIOURS

Most bees need a cavity in which to raise their young; however, over time, the many different species have developed distinct nesting strategies and preferences. These various nests serve to shelter and protect developing bees from weather, predators, and harmful microbes.

Ground
The vast majority of bee species nest in the ground. Some have special adaptations that allow them to dig, while others prefer to inhabit existing holes, dug by other creatures who no longer need them.

Plant Matter

Some bees prefer to nest in the hollow branches of dead plant matter. A dried stalk serves as a pre-made tunnel for nesting mothers.

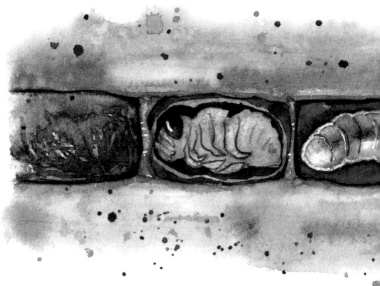

Cross-section of a linear leafcutter bee nest

Wood

Other bees nest in decaying wood. They chew tunnels into tree stumps or even human-made structures.

Miscellaneous

Social bees require more space and will nest in a variety of cavities, from abandoned rodent burrows and hollow logs to more creative spots such as compost bins and barbecue grills.

In temperate climates, honey bees sometimes build open-air hives and use their bodies to heat their nest.

IDENTIFYING BEES

A bee's most recognizable trait is its cuteness. Bees are often confused with flies and wasps, especially because all three visit flowers. Yet bees can be distinguished from these less attractive cousins by several charming qualities.

ANTENNAE

Bees and wasps both have long, elbowed antennae. They sprout from between the eyes and sweep out elegantly from a subtle crook. Yet bees tend to have antennae that move with curiosity and purpose, while the antennae of wasps are often restless and engaged in a constant jittery search for prey. By comparison, flies have short, stubby, unimpressive antennae.

EYES

Few people realize it, but the eyes of a bee are lovely. They are large and pleasingly curved around the sides of the bee's face. Many are a deep chocolate colour, but some species have brightly coloured eyes, in marbled tones of wintry blues, emerald greens, and terracotta reds.

The eyes of a wasp bear a close resemblance to those of bees – attractive and colourful – but without the handsome framing of fuzz, they just look menacing. The eyes of a fly, on the other hand, are large and bulbous. They crowd the fly's head, leaving little room for any other features. You might catch sight of a fly that is fuzzy and round, it may even have stripes like a bee, but the disguise is always spoiled by a pair of bulging eyes.

Close-up of an anthophorine bee (*Anthophora urbana*) eye

WINGS

Wings can be helpful in telling insects apart. The wings of a bee have an almost butterfly shape when open, but fold in elegant layers when not in use. In contrast, flies have only one set of crude, splayed wings that typically do not fold and wasps have long, thin, paddle-like wings.

POLLEN PANTS

When collecting pollen, bees mix the pollen with nectar to make it sticky and pliable, then pack it onto their back legs. Depending on the type of bee, the pollen will either take the shape of a large ball or create a thick layer over the entire leg. The effect is that the bee appears to be wearing a pair of trousers made of pollen. These adorable 'pollen pants', or trousers, are unique to bees. You will never see a wasp or fly wearing them.

HAIRY AND ROUND

A bee's most endearing trait is its fuzziness. Its legs, belly, and even its eyes are downy. They have special branched hairs that, when examined under a microscope, look more like feathers and are excellent for trapping pollen. By contrast, wasps are often sleek and hairless, while most flies have only a sparse covering of hair that gives them an untidy appearance.

Bees also tend to have thick, stout bodies that are very different from the lean and slender bodies of most wasps. Their legs, too, are more thickset and shapely than those of the wasp. Even the species of bee that are slim-bodied have a roundness to them that contrasts with the wasp's angular lines.

BEE FACT

AS WITH MOST ANIMALS, MALE
AND FEMALE BEES OFTEN HAVE
A DIFFERENT APPEARANCE.
DEPENDING ON THE SPECIES, MALES
MAY HAVE DISTINCT COLOURING OR
A LARGER OR SMALLER BODY WHEN
COMPARED WITH THE FEMALE.

Female metallic green sweat bee

Male metallic green sweat bee

YET MALE BEES DO SHARE SOME
COMMON TRAITS ACROSS SPECIES
LINES. FIRST, MALE BEES DON'T
HAVE STINGERS. SECOND, THEY
DON'T HAVE ANY BIOLOGICAL
FEATURES THAT WOULD ALLOW
THEM TO COLLECT POLLEN, SUCH
AS SCOPAE OR CORBICULAE. THIS
IS BECAUSE THEY DON'T COLLECT
POLLEN TO FEED THEIR YOUNG.

Bee mimic masquerading as a bee

BEHAVIOUR

Bees hardly ever leave their work to bother humans. Wasps, on the other hand, have a bad reputation for ruining picnics – a reputation shared by some pesky flies. Adult wasps feed on nectar and pollen, but must also hunt meat for their young. This means that wasps might take an interest in both your can of fizzy drink *and* your ham sandwich! And while some bees may attempt to share your fizzy drink with you during times of nectar dearth, they'll never take a bite of your sandwich.

BEE MIMICS

There are an awful lot of insects that look like bees but aren't actually bees. These copycats are called mimics and they are so good at what they do, they often fool us! The bee's sting is so envied by other insects that many have evolved to look like a bee in the hope that their bee-like appearance will ward off predators that fear stings. Among the most common of these mimics is the flower fly, which disguises itself as a bee or wasp with bee-like fuzz or black-and-yellow markings.

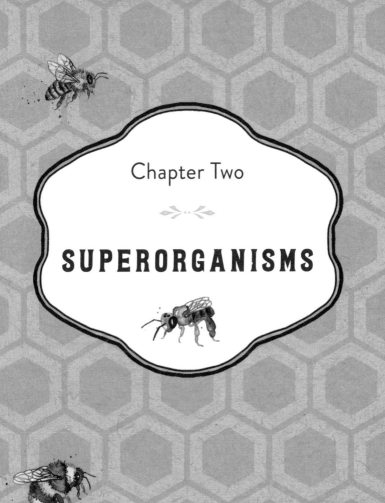

Chapter Two

SUPERORGANISMS

SOCIALITY IN BEES

Social bees have given up the solitary life to live in a group. There are varying levels of sociality among bee species. Some bees live in aggregations: they maintain their own nests and raise their own young, but do so in close proximity to other bees. Other species, called communal bees, go one step further and share a nest entrance but keep to their own separate chambers once inside. Think of it like a bee apartment building.

Truly social bees, or eusocial bees, share a single nest and divide up the work of maintaining it. They live in family groups made up of a single mother, the queen, and her many daughters, the workers. Depending on the species, these colonies can grow quite large; some may have as many as 100,000 members, all working together in a form of complex cooperation, as if they were one animal. Many people assume that all bees live this way, but social bees actually account for only a small percentage of bee species. Only honey bees, bumble bees, stingless bees, and some sweat bees are eusocial.

BUMBLE BEES

These much-loved, larger-than-average bees are easily recognized by their trademark hum and profusely fuzzy bodies. Even though there are some 250 species worldwide, many of them look similar to one another. Most are plump in shape with broad white, red, or yellow stripes. They are frequenters of gardens and often choose to nest in common garden features, such as compost heaps, woodpiles, rockeries, and nest boxes. Yet most bumble bees nest in the ground and go unnoticed because their colonies do not grow large. A single bumble bee nest typically houses fewer than 200 bees.

THE BUMBLE BEE ANNUAL CYCLE

A bumble bee colony begins with the death of the one before it. This is why bumble bee nests are modestly sized, because they must start again each year.

In late summer, the colony's last effort is to raise queens. If successful, these queens will establish new colonies the following spring. A young queen's first task is to leave the nest to mate. Male bumble bees are often found waiting in groups near flower patches, ready to compete with each other, should a queen happen by. After mating, the male bumble bee dies, but if the queen survives the winter, his genes will live on in the new colony. The queen gorges on the last of the summer flowers, fattening herself up before she burrows underground where she will overwinter.

BEE FACT

BUMBLE BEES ARE AMONG THE
FIRST BEES SEEN IN SPRING AND
THE LAST BEES OUT IN AUTUMN
BECAUSE THEIR SPECIAL
SHIVERING ADAPTATION
ALLOWS THEM TO FLY IN COOLER
TEMPERATURES THAN MOST
BEES. THEY CAN RAISE THEIR
INTERNAL TEMPERATURE TO
30°C (86°F) USING THIS METHOD,
EVEN IN TEMPERATURES
AS LOW AS 7°C (45°F).

In early spring, the lone, hibernating bumble bee queen is one of the first bees to emerge. The hungry queen gorges herself on the nectar and pollen of early flowers such as lilac, heather, and comfrey. Then she begins the work of establishing her nest. Most bumble bee queens choose abandoned rodent burrows or similarly sheltered spaces to build in. The queen then collects soft materials such as feathers or moss to arrange into a loose, round structure that she can roost in. Bumble bee queens must sit on their eggs to keep them warm, exactly like a mother bird.

The queen collects pollen and shapes it into a sticky ball, before pressing her eggs into an indent at the top. She then seals it with wax, positions herself on top and starts the work of keeping it warm. To raise her core temperature, she uses a special adaptation called shivering, where she vibrates her flight muscles to generate heat. During the two-week incubation period, the queen must keep her brood about 30°C (86°F) and cannot leave them for very long. She prepares herself for this by stockpiling nectar ahead of time in little wax cups that she arranges around her eggs. Inside the sealed pollen ball, her eggs hatch into larvae and begin to eat the pollen. Once they are large enough, they spin cocoons and pupate. When they emerge as adults, they will be the colony's first worker bees.

Worker bees are all female and perform a range of tasks, but the first group starts by taking over the responsibility of warming the next batch of young. With these new helpers in the nest, the queen is free to leave on foraging trips. These ventures out of the nest will be her last, though; once the workers have grown strong enough, they will take over the work of food collection.

As the population of worker bees grows, the queen is able to lay more and more eggs, which will hatch into more worker bees. The colony reaches its peak size in midsummer, with dozens or sometimes hundreds of worker bees living together with a single queen. At this time, the colony begins to produce male bees and, soon after, they will make next year's queens. Most male bumble bees never successfully mate, but they dutifully leave the nest in an attempt to do so and do not return.

By autumn, the bumble bee colonies have reached their end. Only a fraction of the queens from the previous year succeed in growing their colonies through summer. Many don't survive hibernation or cannot find a nest. Some starve while trying to warm their eggs. Others are eaten by badgers, killed by disease, or poisoned by pesticides. And so, when a colony makes it to autumn and its new queens have left the nest, it's not a sad occasion, but a triumphant one.

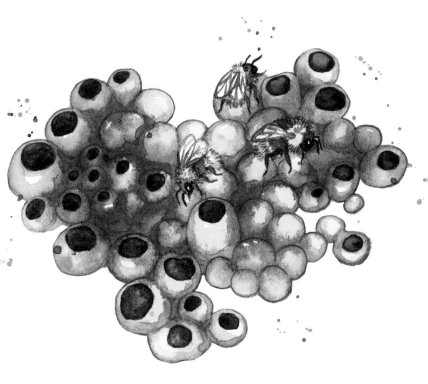

A mature bumble bee colony with workers

Common carder bee

Bombus dahlbomii

A common carder bee shown here to scale next to the *Bombus dahlbomii*

THE WORLD'S LARGEST BUMBLE BEE, *BOMBUS DAHLBOMII*, IS RUMOURED TO BE 4 CM (1½ IN) LONG AND, WHEN AIRBORNE, RESEMBLES A FLYING MOUSE. THE GIANT, FLUFFY ORANGE BEE, ONCE COMMON IN ITS NATIVE LAND OF CHILE, IS NOW FEARED TO BE EXTINCT. IT'S BELIEVED THAT THE NON-NATIVE BUFF-TAILED BUMBLE BEES IMPORTED BY FARMERS FOR POLLINATION CARRIED A DISEASE THAT MAY HAVE WIPED OUT *BOMBUS DAHLBOMII* ENTIRELY.

BUMBLE BEES OF EUROPE

Europe has the most diverse bumble bee population, with more than 68 species living throughout the continent.

Common carder bee
(*Bombus pascuorum*)
10 mm to 15 mm (⅜ in to ⅝ in)

White-tailed bumble bee
(*Bombus lucorum*)
12 mm to 18 mm (½ in to ¹¹⁄₁₆ in)

Early bumble bee
(*Bombus pratorum*)
10 mm to 14 mm (⅜ in to ⁹⁄₁₆ in)

Red-tailed bumble bee
(*Bombus lapidarius*)
11 mm to 16 mm (⁷⁄₁₆ in to ⅝ in)

Buff-tailed bumble bee
(*Bombus terrestris*)
11 mm to 17 mm (⁷⁄₁₆ in to ¹¹⁄₁₆ in)

Garden bumble bee
(*Bombus hortorum*)
19 mm to 22 mm (¾ in to ⅞ in)

Tree bumble bee
(*Bombus hypnorum*)
9 mm to 15 mm (¹¹⁄₃₂ in to ⅝ in)

Gypsy's cuckoo bumble bee
(*Bombus bohemicus*)
15 mm to 20 mm (⅝ in to ²⁵⁄₃₂ in)

BUMBLE BEES OF NORTH AMERICA

North America boasts almost 50 bumble bee species. Many of these have a large range, but are limited to one side of the Rocky Mountains or the other.

Common eastern bumble bee
(*Bombus impatiens*)
9 mm to 16 mm ($^{11}/_{32}$ in to $^5/_8$ in)

**Tri-coloured or
orange-belted bumble bee**
(*Bombus ternarius*)
8 mm to 13 mm ($^5/_{16}$ in to ½ in)

Red-belted bumble bee
(*Bombus rufocinctus*)
9 mm to 13 mm ($^{11}/_{32}$ in to ½ in)

**Yellow-faced or
'Vosnesensky' bumble bee**
(*Bombus vosnesenskii*)
8 mm to 17 mm ($^5/_{16}$ in to $^{11}/_{16}$ in)

Black-tailed bumble bee
(Bombus melanopygus)
11 mm to 15 mm (⁷⁄₁₆ in to ⁵⁄₈ in)

Hunt's bumble bee
(Bombus huntii)
11 mm to 14 mm (⁷⁄₁₆ in to ⁹⁄₁₆ in)

Two-form bumble bee
(Bombus bifarius)
8 mm to 14 mm (⁵⁄₁₆ in to ⁹⁄₁₆ in)

Yellow bumble bee
(Bombus distinguendus)
11 mm to 17 mm (⁷⁄₁₆ in to ¹¹⁄₁₆ in)

BEE FACT

BUMBLE BEES HAVE DEMONSTRATED AN IMPRESSIVE ABILITY TO LEARN COMPLEX TASKS AND SOLVE PROBLEMS. THEY EVEN HAVE THE ABILITY TO LEARN FROM EACH OTHER. SCIENTISTS AT QUEEN MARY UNIVERSITY OF LONDON SUCCESSFULLY TRAINED BUMBLE BEES TO TUG ON A STRING, AND IN ANOTHER EXPERIMENT, TO ROLL A SMALL BALL INTO A PAINTED CIRCLE FOR SUGARY TREATS.

THEY OBSERVED THAT BEES
WHO WATCHED AN ALREADY
TRAINED BEE PERFORM
THESE TASKS WERE ABLE TO
PICK UP THE SKILL FASTER.

HONEY BEES

Honey bees have the most highly evolved social structure of all bees. They live in large, complex groups, whose members are capable of sophisticated communication. Like bumble bees, their colonies have a single egg-laying queen, a body of female workers, and a small percentage of males during mating season. Unlike bumble bees, honey bee colonies are enduring; they don't die at the end of summer. In fact, a honey bee queen can live for several years and a colony can live for decades.

THE SWARM

Honey bee colonies begin in spring when a portion of an existing colony leaves with the queen to establish a new nest. The newly formed colony, or swarm, flies together in a massive cloud with a buzz so loud, it briefly blocks out all other sounds. Although it may look alarming, the swarming bees are not interested in attacking anyone, they are only travellers on their way to a new home.

BUILDING COMB

When a swarm arrives at its chosen nest site, the bees arrange themselves in a cluster that hangs down from the top of the cavity. The mass of bees is actually made up of long, interconnecting chains of worker bees who hang onto one another's legs like small circus performers. Collectively, they maintain a core temperature of 36°C (97°F), which is crucial to the construction of the nest architecture: comb.

Comb is the impressive beeswax structure, made up of hexagonal cells, in which bees raise their young and store their food. To build it, they mould small flakes of wax excreted from their abdominal glands. The bees construct several pieces of comb with small gaps in between. Outer combs are dedicated entirely to honey storage, while inner combs are used to rear the brood.

HONEY BEE LIFE CYCLE

Inside the swarm cluster, the queen bee begins to lay eggs destined to become worker bees. She deposits a single pearl-coloured egg at the bottom of each hexagonal cell. Each tiny egg hatches into a larva that must be fed by the worker bees. In its first few days, the larva consumes a special

LIFE CYCLE OF THE HONEY BEE

Queen
lays egg

Larva
day 6

enzyme-rich substance called royal jelly – the equivalent of mother's milk. Then it is switched to a diet of pollen and honey until it grows large enough to pupate. When a larva is ready to transition into its adult form, worker bees stop feeding it and cover the cell with a beeswax cap. When the new bee emerges, still downy with a soft, white fuzz, she sets to work immediately.

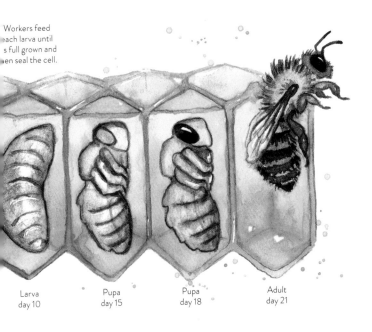

Workers feed each larva until s full grown and en seal the cell.

Larva
day 10

Pupa
day 15

Pupa
day 18

Adult
day 21

WORKER BEE JOBS

Honey bee colonies average 50,000 bees, the majority of them female worker bees. Each worker goes through a series of jobs that change as they age. When they are young, they perform tasks inside the nest, such as cleaning, building comb, or nursing larvae. As they mature, they graduate to more risky assignments such as guarding the entrance or foraging for food.

Eventually, all worker bees become foragers and they perform this task until they die. They will travel up to 5 km (3 miles) from their nest to collect food, some making up to 15 trips a day. You can recognize particularly old worker bees by their haggard, chipped wings and sparsely haired bodies – older bees tend to go bald from all the hard work they've done.

HONEY BEE COMMUNICATION

Honey bees communicate with each other in several complex ways. One of the most prevalent methods is scent. The beehive has a wonderful perfume: to humans, the warm air smells like honey and freshly baked bread. The bee's sense of smell is 100 times more powerful than ours and it produces a range of chemical scent signals, called pheromones, that our noses can't always detect. Pheromones are involved in nearly every aspect of honey bee life – brood rearing, foraging, navigation, defence, swarming, and mating – it's how they organize themselves.

Pheromones are often coupled with other forms of communication, such as vibratory signals. We all know that honey bees buzz, but what most people don't realize is that they are not able to hear the sounds they make; instead, they *feel* it. Honey bees send messages through the comb by vibrating their bodies. The comb acts as a conduit for these communications.

The most famous form of honey bee communication is dance. Forager bees use dance primarily to communicate the location of high-quality floral sources. The waggle dance is a sequence of circular movements that resembles the shape of a figure of eight. The dancing bee walks in a straight line, sashaying her abdomen from side to side as she goes, then turns in a circular motion, repeats her waggling, and turns again. The direction she faces during the waggle portion of her dance in relation to the sun communicates an angle that tells the other bees how to get to the flowers. A bee that dances in a straight line to the left creates a 90-degree angle. That means the flowers she's found are 90 degrees to the left of the sun.

THE WAGGLE DANCE

Sun

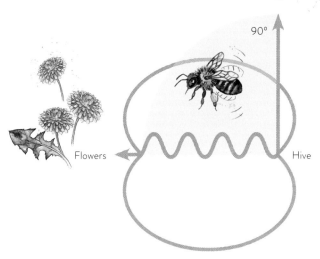

90°

Flowers

Hive

110°

60°

THE MATING FLIGHT

Even though she can live for up to seven years, the queen bee mates only once. Soon after emerging from her cell as an adult, she embarks on her mating flight to find male bees, called drones, from other colonies. These large, fuzzy males have enormous eyes for sighting eligible queens. Every day drones leave the hive hoping to meet a mate. They hang out with males from other colonies in the same locations year after year, although we still don't understand how they choose these mysterious sites. When a virgin queen flies by, the drones race to mate with her, forming a comet behind her as they jockey for position. The queen will mate with as many as 20 males on her flight. She stores a little semen from each one in a special organ, called the spermatheca, and draws from this supply for the rest of her life. Unfortunately for them, the males who successfully mate do not live to tell the tale. Drones die during the act or directly after.

MOST QUEEN BEES HAVE TO FIGHT FOR THEIR TITLE. WHEN A COLONY NEEDS A NEW QUEEN, THEY OFTEN MAKE MORE THAN ONE, AND SOMETIMES AS MANY AS 20! THESE NEWLY EMERGED QUEENS MUST THEN BATTLE TO THE DEATH, VICIOUSLY BITING AND TEARING AT ONE ANOTHER IN AN ATTEMPT TO DELIVER A FATAL STING, WITH THE VICTOR GOING ON TO MATE AND TAKE ON THE RESPONSIBILITIES OF A QUEEN.

BEE FACT

HONEY BEES CAN RECOGNIZE
INDIVIDUAL HUMAN FACES.
THOUGH THEIR BRAINS ARE
TINY, THEY ARE CAPABLE OF THE
SAME COMPLEX COGNITION
THAT ALLOWS US TO PICK OUR
FRIENDS OUT OF A CROWD.

STRANGE GENETICS

When a queen lays an egg, she can choose to lay a fertilized or an unfertilized egg. Unfertilized eggs are male and will develop into drones. Males are produced only in late spring and summer when virgin queens from other colonies may be looking for a mate. Fertilized eggs are female and have the potential to become either a worker or a queen depending on what they are fed. Most will become worker bees, but their destiny depends on the needs of the colony.

If a new queen is required to replace an aging or ailing one, worker bees can make one from a fertilized egg. Queen bees are too large to develop in a regular cell, so workers begin by building a special queen cell. The enlarged chamber is roughly the same size as a peanut. Inside, the royal larva is given a generous supply of royal jelly to feed on. Unlike worker bees and drones, the queen larva is never fed pollen or honey. This is what allows her to develop into a queen bee. She may serve her colony for many years, laying thousands of eggs in her lifetime.

THE INFINITE COLONY

At the end of summer, honey bee colonies begin to prepare for winter. Over the last few months, the bees have stored a tremendous amount of food. A single colony can make as much as 45 kg (100 lb) of honey in a season. This is vital to their winter survival. Although they cannot fly in temperatures below 10°C (50°F), honey bees do not hibernate over the winter months. Instead, they cluster tightly together and consume large amounts of honey to keep themselves warm. In temperate climates, honey bees can fly and forage year-round, but they still rely on stored honey from the spring during the winter months.

If a colony survives winter, it will continue the cycle of reproduction that spring when it sends out a swarm of its own. After the old queen leaves with the swarm, a new queen is raised to replace her and the colony continues. If it weren't for predators, disease, and extreme weather events, a honey bee colony might live on forever.

BEE FACT

HONEY BEES ARE AMONG THE
ANIMALS THAT HAVE ACCOMPANIED
US TO SPACE. TWO SMALL COLONIES
JOINED THE 1984 *CHALLENGER*
MISSION. UNLIKE FLIES AND
MOTHS WHO HAD DIFFICULTY
FLYING IN MICROGRAVITY, HONEY
BEES INITIALLY STRUGGLED, BUT
AFTER SEVERAL DAYS THEY HAD
LEARNED TO FLY AND EVEN
MANAGED TO BUILD COMB.

Stingless bee hive honey pots

STINGLESS BEES

Stingless bees, also called Meliponines, are tropical bees, with about 500 different species worldwide. They are native to Central and South America, Southeast Asia, Africa, and Australia. Stingless bees are closely related to honey bees and share many common traits with them, including the production of honey, but they cannot sting. Instead, they defend their hives by biting, although most species are considered harmless to humans.

NESTING

Stingless bee colonies can grow quite large, and range from a few hundred bees to as many as 80,000 individuals. They often nest in tree hollows, but will also take up residence in less conventional cavities such as water meter boxes, bins, and even inside the walls of some structures. They build horizontal combs from a mix of wax and plant resins. These structures are divided into two distinct sections: a central comb structure used to raise young and an outer food storage area made up of small pots that contain honey and pollen. One Australian species builds beautiful layered brood combs that take on the shape of a spiral.

COLONY LIFE

Like other social bees, stingless bee colonies are organized by a caste system, and the population is divided into queens, workers, and drones. A colony may have two or more queens laying eggs in the same nest. It's also not uncommon for a colony to raise extra queens, who will remain virgins until they might be needed.

After a laying queen deposits an egg in a brood cell, the worker bees are not responsible for feeding the growing larvae. Instead, each cell is provisioned with a supply of honey and pollen before the egg is placed in it and then it's sealed up. Inside, the growing larva has all the food it needs to develop. Queens also receive pollen and honey during their larval stage; they just get more of it and that appears to be what triggers them to develop into queens. The majority of the brood raised will become female worker bees, but the queen also lays unfertilized eggs to produce the male drones. Since they are residents of warm, tropical climates, stingless bees are active year-round. Their colonies can last for decades and their queens live anywhere from two to seven years.

Stingless bee queen

COMMUNICATION

Stingless bees communicate using scent, vibratory signals, and dance. Depending on the species, stingless bee foragers have developed different methods for communicating information about potential food sources, but unfortunately, they are not well understood. In some species, returning foragers lead nestmates to rich floral sources by creating a scent trail – a unique behaviour not shared with their honey bee cousins. These bees mark rocks and tree trunks between their nest and the forage site to help guide other foragers. Other species use pulse signals and dance to communicate the height and distance of potential food sources. They emit high-pitched noises accompanied by vibrations as they turn in circles or zigzag for an audience of nestmates.

THE SWARM

Stingless bees create new colonies by swarming, but their process is much more careful and gradual than that of the honey bee. They don't take flight in a large group with their queen. Instead, a small team of workers transports building materials to the new nest site. Over the next few days or weeks, more and more workers will move from the mother colony to the new nest site and eventually a queen joins them and begins egg production. If the new colony is successful, it will go on to establish its own daughter colonies.

Chapter Three

HONEY

Honey has been prized by humans for thousands of years for its sweetness, nutrition, and medicinal properties. The collection of honey is one of the oldest known human activities and remains popular today – despite the risk of stings.

WHAT IS HONEY?

Honey is concentrated nectar. Bees use their tongues to suck the nectar from flowers and carry it back to their nest inside a special organ, called a honey gut. Back at the hive, it's deposited into honeycomb and dehydrated until it becomes a thick, sweet syrup. Honey is one of the only true natural sweeteners and contains nutrients, enzymes, minerals, antioxidants, and amino acids. Yet most bees don't make honey, or not a significant amount of it. Only honey bees and some stingless bees produce enough honey for humans to share in.

TYPES OF HONEY

The flavour, consistency, colour, smell, and medicinal properties of honey are dependent on the type of flower from which the honey is made. In most settings, bees collect nectar from a variety of sources and these nectars combine to create a unique flavour profile that is different with every harvest.

Depending on the location and time of year, there may be a dominant bloom that honey bees favour and so their honey will be largely influenced by the properties of that nectar source. Beekeepers in different regions have learned to recognize local nectar sources and will label their honey as such. Examples include buckwheat honey, sage honey, tupelo honey, eucalyptus honey, and heather honey.

LIKE MOST ANIMALS, BEES NEED WATER TO SURVIVE. HOWEVER, HONEY BEES DON'T JUST DRINK WATER – THEY USE IT TO COOL THEIR HIVES. WHEN THE TEMPERATURE RISES, FORAGERS CARRY WATER BACK TO THEIR HIVE USING THEIR HONEY GUTS. INSIDE, THE BEES SPREAD THE WATER ON THE COMBS AND FAN THEIR WINGS TO LOWER THE TEMPERATURE THROUGH EVAPORATIVE COOLING. IN EXTREME HEAT, SOME BEES TRANSFORM INTO LIVING WATER

TANKS, USING THEIR BODIES TO STORE THE PRECIOUS LIQUID AND DISPENSING IT AS NEEDED TO OTHER MEMBERS OF THE HIVE.

Other beekeepers can be more confident still in how they label their honey because they keep their bees on farms where only one type of crop is grown. In these vast agricultural settings, the bees are bound to make 'orange' honey or 'avocado' honey, for example, simply because nothing else is growing. However, while it may sound appealing to consumers, most of the bees who yield varietal honey suffer undue stress from poor nutrition and exposure to pesticides.

The best way to buy honey is direct from a local beekeeper. Look for raw (unheated), minimally filtered honey and avoid agricultural honeys made from a single crop. Honey is healthier for bees and humans when it comes from a more natural setting, with many flowers available to choose from. These honeys are sometimes called 'wildflower'.

PRACTICAL USES FOR HONEY

There are so many ways to enjoy honey – chances are, you aren't using it to its full potential!

BAKING

Honey is more healthy than processed sugar and can be used to create more wholesome recipes. It's also an opportunity to add a unique, new flavour and it helps keep baked goods moist. When replacing sugar with honey in a recipe there are a few basic guidelines to follow. First, use less honey than sugar. The exact amount should depend on the sweetness of the honey you are using, but typically you want to use 170–225 g (½–⅔ cup) of honey for every 225 g (1 cup) of sugar. Second, you'll want to reduce the other liquids in your recipe. Unlike dry sugar, honey can change the consistency of your recipe, so to make up for it, try reducing other liquid ingredients by 60 ml (¼ cup) for every 340 g (1 cup) of honey added. Lastly, you should reduce your oven temperature by 10°C (25°F); honey can burn easily, so keep a watchful eye on it. You can also use honey in frosting or as a drizzle for baked goods.

COOKING

Honey is the perfect ingredient for cooks who like to experiment. With so many varieties available, there are endless flavours and textures to cook with. To start, try using honey in glazes, marinades, salad dressings, or sauces. In some cases, honey by itself is a wonderful addition to a meal. Drizzle it over sautéed carrots or on your next slice of pizza.

SWEET DRINKS

Try using honey instead of sugar in tea, coffee, lemonade, smoothies, and cocktails. Not only is it better for you, but it creates a new dimension of flavour, depending on the variety of honey you use. It can make your drink more citrusy or lend a floral element, or it may add a deep molasses-like tang. Honey can also mellow strong flavours, resulting in a smooth taste that works especially well for cocktails. When adding honey to cocktails or other cold drinks, dissolve it in a small amount of warm water before adding it to your recipe for best results.

HANGOVERS

If you happen to have had one too many honey-sweetened cocktails, you might be surprised to learn that honey is also an effective hangover cure. Having a spoonful of honey or adding a generous drizzle to your morning toast may relieve some of your hangover symptoms. The fructose in honey helps the body break down and process alcohol more quickly. It can also give you energy that will keep you from napping the day away.

A NATURAL ENERGY BOOST

When you are feeling fatigued, honey can provide a long-lasting energy boost. Processed sugars contain glucose that gives the body an immediate energy boost, swiftly followed by a crash, but fructose is absorbed more slowly and provides sustained vitality. Lucky for us, honey has both. So when you consume honey, you can expect a swift and stable energy boost. If you've planned a long day of activity with children, like a trip to the zoo, try bringing a small jar of honey or some honey sticks with you as a snack. You'll be amazed at how quickly it revitalizes tired children. Honey is also a wonderful option for pregnant women when they go into labour. It can give them the sustained strength they need to deliver their baby.

BEE FACT

ARCHAEOLOGISTS HAVE UNCOVERED
HONEY BURIED IN ANCIENT
EGYPTIAN TOMBS, DATING BACK 3,000
YEARS, THAT IS STILL EDIBLE TODAY.

HONEY IS THE ONLY FOOD ON
EARTH THAT DOES NOT SPOIL. THIS
IS BECAUSE THE WATER CONTENT IN
HONEY IS SO LOW THAT BACTERIA
CANNOT GROW IN IT. MANY
CONSUMERS MISTAKENLY BELIEVE
THAT WHEN HONEY CRYSTALLIZES,
IT HAS GONE BAD. IN REALITY,
CRYSTALLIZATION IS A NATURAL
PROCESS THAT CHANGES ONLY

THE TEXTURE OF THE HONEY. AS
LONG AS HONEY REMAINS SEALED
IN ITS JAR, IT CAN LAST FOREVER.

SKINCARE

Raw honey is excellent for skincare, and it's especially useful for managing acne. Its anti-inflammatory properties soothe irritated skin and reduce redness. It's also antibacterial, so it can resolve and prevent infection. Lastly, it's extremely moisturizing so it keeps your skin healthy and clear. You can use honey as a spot treatment, applying it directly when problems arise, or as a nightly facial mask, or you might like to add it to a bath for a full-body benefit.

HAIRCARE

If your hair is dry or damaged, honey can help, by sealing in moisture that prevents future breakage. It can also resolve dry scalp issues, such as dandruff, and help hair grow more thickly when applied to the scalp because it strengthens hair follicles. Create a honey mask by mixing honey with warm water, olive oil, coconut oil, or any other moisturizing base. Then apply the honey mixture to your scalp and work it through your hair. Let the mask set for 30–60 minutes before rinsing out.

HONEY HEALTHCARE

Once a staple of ancient healthcare, honey has since been dismissed as a folk remedy, but is now making a comeback in modern medicine. Although more research needs to be done, honey has shown promising medical benefits largely because of its antioxidant, antibacterial, and anti-inflammatory properties. Results have been especially good for wound care, cough suppression, and allergy relief.

WOUND CARE AND BURNS

Honey can be used topically on cuts and burns with amazing results. Honey is a natural pain-reliever. It's particularly soothing on kitchen burns and sunburns. Not only that, but it may actually speed up the healing process and reduce scarring. Honey also has antibacterial properties that can prevent and, in some cases, resolve infection.

COUGH AND SORE THROAT

The next time you get a cold, ditch the cough syrup and reach for some honey instead. Honey soothes sore throats, reduces coughing, and relieves pain. It also has antiviral properties that could have you feeling better faster. It's common to add honey to tea when sick, but a spoonful by itself works best.

ALLERGIES

If you are allergic to seasonal pollen, local honey may help alleviate your symptoms. It is an imprecise treatment, but many have found relief just by eating a spoonful of local honey each day. The theory is that local honey likely contains small amounts of the pollen you are allergic to, and by consuming it, you can build up an immunity to the allergen. For that reason, allergy sufferers must seek out honey that's local to them, ideally made in the season when their allergies are at their worst.

WHAT TYPE OF HONEY SHOULD I USE?

Manuka Honey

Manuka honey is made predominantly from the nectar of the manuka tree or tea tree, native to New Zealand and Australia. Although most honey has some medicinal value, manuka honey has elevated antibacterial properties that set it apart from other honeys. It is highly valued for its effectiveness in treating wounds, skin problems, cold symptoms, and more.

Melipona Honey

The honey from stingless bees, or Melipona honey, is revered in cultures where these species are native. It has been used as medicine by ancient peoples and was especially important in Mayan culture. In some countries, Melipona honey is still used today and is particularly effective in treating eye conditions, such as cataracts. Melipona honey has a higher water content than that of European honey bees, but it does not ferment. This suggests that Melipona honey may have elevated antibiotic activity. Although more research needs to be done, some believe that Melipona honey has unique and superior medicinal properties.

Raw Honey

The medicinal importance of honey is well documented throughout history and is not limited to a particular type. However, some modern honey processing techniques involve heating the honey, which may diminish medicinal benefits. The antimicrobial activity in most honey is connected to the levels of the hydrogen peroxide that is produced by enzymes in the honey. When honey is heated, enzymes and other antibacterial components can be destroyed. For that reason, consumers should look for raw (unheated) honey.

BEE FACT

MANUKA HONEY IS MADE FROM THE NECTAR OF *LEPTOSPERMUM SCOPARIUM* FLOWERS, COMMONLY KNOWN AS 'MANUKA MYRTLE' OR 'TEA TREE'. ALTHOUGH THIS TREE CAN BE GROWN IN MANY REGIONS, TRUE MANUKA HONEY CAN BE PRODUCED ONLY WHEN THE BEES ARE PLACED IN AREAS WHERE IT GROWS IN ABUNDANCE, SUCH AS IN THE WILDS OF NEW ZEALAND. MANUKA HONEY IS SO VALUABLE THAT SOME BEEKEEPERS GO TO EXTREME LENGTHS TO PRODUCE IT.

IN FACT, IT'S NOT UNCOMMON FOR A BEEKEEPER TO HIRE A HELICOPTER TO AIRLIFT THEIR HIVES INTO REMOTE REGIONS RICH IN MANUKA TREES.

Chapter Four

BEEKEEPING

WHY KEEP BEES?

There are so many wonderful motivations for pursuing the hobby of beekeeping – the creation of honey, aiding pollination, and a simple love of nature – but after gaining some experience, many decide beekeeping is not for them. Between the bee stings, high learning curve, and costly equipment, only a fraction of new beekeepers will make it past their second year. Those who do, often find that they are no longer in it for the honey or pollination, but for the love of honey bees.

AN INTRODUCTION TO BEEKEEPING

BEE-CENTRIC BEEKEEPING

As a new beekeeper, it's easy to forget that you are about to take on the responsibility of caring for several thousand new pets. On your beekeeping journey, you will face many choices. Many large-scale beekeeping practices favour techniques that make the beekeeper's life easier, despite the consequences for their bees, but there is a growing movement of 'bee-centric' beekeepers who strive to put their bees first.

New beekeepers can best serve their bees by learning as much as they can about their fuzzy friends. Read honey bee biology books as well as material on beekeeping techniques. Take the time to observe your bees and how they react to you and your beekeeping methods. When you gain an in-depth understanding of the inner workings of your hive, you can make better choices for your bees.

SELECTING A HIVE STYLE

Beekeepers often develop a preference for a certain hive style as they become more experienced and have the opportunity to try different designs, but many new beekeepers start with the classic Langstroth hive.

A Langstroth hive is made up of a series of boxes that each house a single colony. Inside each box are ten wooden frames for the bees to build their comb in. The framed comb can then be lifted out and inspected or moved from one box to another. And with so many experienced beekeepers working Langstroth hives, for the beginner there is no shortage of information on beekeeping techniques that pertain to this design. Its only real downside is that it involves heavy lifting.

Top-bar hives have a horizontal instead of a vertical design. This means that you will never have to lift a box full of heavy honey, only single bars from which the comb hangs. The simplistic design is meant to mimic a natural log cavity and is thought to have originated in Africa.

Warré hives combine design elements from both the Langstroth and the top-bar. They utilize the same bar style as top-bar hives, but in a vertical design made up of stacked boxes, albeit smaller than the Langstroth hive.

The Flow Hive is a modified Langstroth hive that allows beekeepers to harvest their honey through a tap on the side of the box. When the honey is ready, the specialized plastic honey frames channel it out of the hive and directly into jars. Although the hive must still be managed, like any other, the process of harvesting the honey is greatly simplified.

Langstroth hive

EQUIPMENT AND TOOLS

Before you get started on your beekeeping adventure, you'll need three essential items: a bee suit, a smoker, and a hive tool. There are many other fun gadgets to try out in addition to these, but if you want to conserve your budget, you can pass on them.

Bee Suit

When selecting a bee suit, go for a design in which a zip connects the hood to the body. Tie-on hoods, and veils with elastic straps, can easily shift and create a breach in the protection your suit provides. Standard bee suits are made from a thick cotton-canvas, but if you're willing to invest a little more, ventilated suits, made from layers of mesh, are well worth the extra money. These suits allow for air to flow through, while still protecting you from stings.

Smoker

Smoke has been used to calm bees for thousands of years. When a bee colony is disturbed, they mount their defence through the release of a strong odour, similar to that of bananas, called the alarm pheromone. This scent spreads quickly through the hive and triggers bees to act defensively. The strong smell of smoke temporarily blocks this signal and allows the beekeeper to inspect the colony in relative peace.

Hive Tool

A hive tool is a kind of mini crowbar that beekeepers use to pry apart boxes and lift frames during hive inspections. Many new beekeepers are surprised when they try to lift the lid of their newly established colony and find that it is stuck. Honey bees make an antimicrobial substance called propolis, which they use to coat the inside of their hive and fill cracks. It works like a kind of protective glue that has an impressive hold and makes a hive tool an absolute must.

CHOOSING A LOCATION

Beehives aren't just found in the country and on farms – these days, you can find them in a variety of urban settings, from balconies to hotel rooftops. The adaptive nature of honey bees gives us plenty of options when it comes to placing a hive.

Accessibility

Make sure you place your hive in a location where you can work safely and comfortably. Ideally, you should have enough space to stand behind your hive as well as to the side. Remember you are going to be lifting heavy equipment, while sweating in a cumbersome bee suit, under threat of stings! Your location should be flat, stable, and weight-bearing.

Sun

Bees do well in full sun, but if you are in a particularly hot climate, you may opt for partial shade or consider setting up a temporary shade solution, such as an umbrella, during the hottest months. If possible, position your hive so that the entrance receives morning sun.

Wind

Bees prefer to be sheltered from wind. If you're planning your apiary somewhere breezy, consider placing your hive in an area with a natural wind break, or you may need to make your own. Hedges and trees will keep your bees from blowing off course, and they may help insulate them against the cold during winter.

Flight Path

During the day, the space in front of your hive entrance resembles a busy airport, with many bees taking off and landing. This flurry of activity usually stretches 1.5–3 m (5–10 ft) in front of the hive. Take care when placing your bees in populated spaces. Give the colony at least a 3 m (10 ft) radius of space where there is little to no human activity. Face the entrance away from active areas. If you're working with a small or heavily trafficked location, consider elevating the hive by placing it on an accessible balcony or rooftop.

HOW TO GET BEES

The best time to establish a new colony is in spring. Most beginners purchase their first colonies from an experienced local beekeeper. Contact nearby apiaries and beekeeping clubs about purchasing a 'starter colony'. You may need to reserve a colony early in the year because many suppliers sell out as spring approaches. If you're the adventurous type, you could get started by catching or attracting a swarm, but you'll need to do a little research before attempting to do so. See pages 212–213 for a list of additional resources.

INSPECTIONS AND MAINTENANCE

It's important to do regular hive inspections. You should plan to open your hive once every two to three weeks to check the bees' progress and monitor their health, but don't open it any more than that or you risk stressing the colony. During inspections, you should make sure your bees have room to build new comb, assess their honey and pollen levels, confirm you have a queen, and evaluate their brood pattern. To do this you must gain an understanding of what everything looks like in your hive.

When inspecting the brood, remember that bees go through four life stages: egg, larva, pupa, and adult. In a healthy hive, the queen lays a single egg at the bottom of a cell and then does the same in all the surrounding cells, creating a patch of brood that is close in age. When the eggs hatch, the larvae should be pearly white and curled in a C shape. Once they are ready to pupate, the nurse bees cover them in their cells with a beeswax cap. There should now be a large, solid patch of capped brood, and this is what is described as a healthy brood pattern.

In an unhealthy colony, sick larvae are sometimes discoloured, malformed, or twisted in their cells. The beekeeper doesn't always see these afflicted larvae because worker bees quickly remove them from their cells and the queen soon lays an egg to replace them. As a result, the brood nest will have an uneven, patchy pattern; the more irregular the pattern, the worse the problem.

New beekeepers may not be capable of diagnosing exactly what ails their hive, but if they can recognize that there is a problem, they can seek help from experienced beekeepers.

HONEY HARVESTS

As a general rule, beekeepers do not harvest honey from first-year colonies. It's best to let the bees keep what they have made so they can come through the winter strong and healthy. In their second year, you can harvest honey any time, from spring through to autumn. Just make sure you leave enough honey with your bees for winter. It's best to consult with a local beekeeper to find out exactly how much you need to leave for your climate.

When harvesting honey, select frames that consist only of honey. You don't want to harvest honey from frames that also contain brood. These honey frames should be at least 70 per cent capped, which indicates that the honey is mostly ripe and is ready for harvest.

An extractor

Once you have selected the frames you want to harvest, you'll need to brush the bees off them and bring them inside. You then have two options. The first involves a machine called an extractor that allows beekeepers to spin the honey out of the combs using centrifugal force. Many beekeepers favour this method because it allows them to reuse empty comb in their hives. Alternatively, you can cut the entire comb from the frame, harvesting both the wax and the honey.

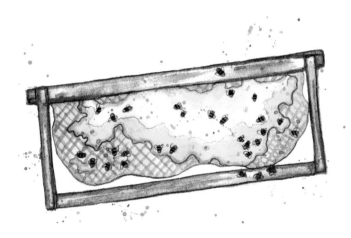

TREATING BEE STINGS

When stings happen (and they will), beekeepers will tell you the best relief initially comes from swearing loudly. After that, each beekeeper has their own favourite salve. No matter which you choose, remember to first remove the stinger. After a bee stings, she dies, but her stinger with venom sac attached will often remain lodged in your skin, and will continue to pump venom into your body for up to 20 minutes. The more venom it delivers, the worse your reaction will be.

Ice: Ice will numb the pain and reduce swelling.

Bicarbonate of Soda Paste: Make a paste of bicarbonate of soda (baking soda) and water or your own saliva and apply it to the sting for fast relief.

Honey: A dab of honey will relieve the pain and reduce swelling.

Lavender Essential Oil: A few drops of lavender essential oil will relieve pain, and reduce swelling and irritation.

Antihistamines: Taking an oral antihistamine may relieve longer-lasting symptoms, such as swelling and tenderness, but it often makes you drowsy.

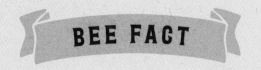

HONEY BEES CAN BE FOUND ON EVERY CONTINENT EXCEPT ANTARCTICA, EVEN IF THEY ARE NOT ORIGINALLY NATIVE TO IT. EUROPEAN HONEY BEES WERE IMPORTED TO NORTH AMERICA IN THE 17TH CENTURY, AND LATER BROUGHT TO AUSTRALIA.

KEEPING STINGLESS BEES

The art of keeping stingless bees, or Meliponiculture, is in decline, but this ancient tradition is worth preserving, and efforts are underway to revitalize Meliponiculture and share best practices.

In places where stingless bees are native, Meliponiculture has fallen out of favour with most beekeepers because their colonies produce very little honey: stingless bees produce between 200 g (7 oz) and 5 kg (11 lb) per colony per year, while a honey bee colony may produce as much as 91 kg (200 lb). In Australia, for example, stingless bees will produce enough honey only for harvest in warmer regions, and although they can be kept in colder climates, they will not survive if the honey is taken. Yet these important native species produce a unique, tangy honey that can be sold for a high price. Additionally, they provide pollination that is often superior to that of the honey bee. They are particularly skilled at pollinating valuable exotic fruits such as avocado, starfruit, macadamia nuts, and mango.

BEE FACT

IN THE 19TH CENTURY, BEEKEEPERS AND THEIR FAMILIES FELT SUCH A CONNECTION TO THEIR HONEY BEE COLONIES THAT THEY KEPT THEM INFORMED OF IMPORTANT LIFE EVENTS IN A CUSTOM KNOWN AS 'THE TELLING OF THE BEES'. THEY WOULD NOTIFY THE BEES OF WEDDINGS, BIRTHS, AND ESPECIALLY DEATHS BY KNOCKING ON THE HIVE AND SPEAKING GENTLY OF THE NEWS. IF THE BEES WERE NOT INFORMED, IT WAS BELIEVED THAT A PENALTY WOULD BE PAID, SUCH AS LOSS OF COLONIES OR POOR HONEY PRODUCTION.

Stingless beekeepers typically keep their bees in traditional log hives or in small hollow boxes that resemble nest boxes, with a very small entrance. Without the danger of stings, stingless beekeepers can place hives in a range of settings. The most important aspect of the stingless beehive is that it remains well sealed to protect it from predators. Stingless beekeepers use mud or tape to seal the cracks between boxes after having broken them during inspections. Some beekeepers install windows on their hive boxes so they can monitor the bees without disturbing them. Honey can be harvested from the small honey pots the bees construct with a syringe, or the pots can be tipped and drained.

BEE FACT

BEEHIVES ARE INCONSPICUOUS AND CAN BE MAINTAINED IN SOME SURPRISING PLACES. IN PARIS, THERE ARE HIVES ON THE ROOFTOPS OF SOME OF THE CITY'S MOST WELL-KNOWN BUILDINGS, SUCH AS THE NOTRE DAME CATHEDRAL. AMAZINGLY, THE NOTRE DAME BEES SURVIVED THE TRAGIC 2019 FIRE THAT DESTROYED THE CATHEDRAL'S ICONIC SPIRE AND CENTRAL ROOF. WHEN FIRE IS NEAR, BEES RETREAT DEEP INTO THE HIVE AND IMBUE THEMSELVES WITH HONEY TO WITHSTAND THE HEAT.

Chapter Five

PROTECTING OUR
BEE BUDDIES

WHY ARE BEES IN DECLINE?

Do you remember stepping daintily with bare feet across your lawn in summer, while bees seemed to buzz between your toes? There was a hum of neighbourhood bumble bees and a nightly orchestra of singing crickets, but lately . . . it's like someone turned the volume down. Researchers from around the world are reporting a dramatic decline in the population of not just bees, but other insects, too. Although a consensus has yet to be reached, most scientists agree that the problem is a combination of pesticides, habitat loss, and climate change. And while we struggle for a definitive answer, some precious species, such as North America's rusty patched bumble bee, are nearing extinction.

BEE FACT

BUMBLE BEE SPECIES IN EUROPE AND NORTH AMERICA ARE EXPERIENCING A RAPID AND DRAMATIC DECLINE IN POPULATION THAT HAS MANY SCIENTISTS WORRIED. A STUDY FROM 2018 EXAMINED 100 YEARS OF BUMBLE BEE RECORDS FOR THE STATE OF VERMONT, USA, AND FOUND THAT NEARLY HALF THE SPECIES LIVING THERE ARE IN SERIOUS DECLINE OR HAVE VANISHED ENTIRELY. IN FACT, FOUR OF VERMONT'S 17 BUMBLE BEE SPECIES HAVE ALREADY BECOME EXTINCT.

SYSTEMIC PESTICIDES

In the mid-1990s a new class of pesticide emerged and since then has gone into widespread global use. These systemic pesticides differ from the insecticides of the past because instead of remaining on the surface of treated foliage, they penetrate the plant's tissues. As a result, systemic pesticides express themselves inside the leaves, roots, flowers, pollen, and nectar of treated plants. After they are applied, they can persist in the environment for months or even years.

A mounting body of research suggests that systemic pesticides are closely linked to the global bee decline. When a bee is exposed to these types of pesticide through floral sources, they don't die immediately, but are negatively impacted in several ways that may lead to their eventual demise. First, their immune system becomes compromised, making them more vulnerable to parasites, disease, and other stressors such as habitat loss and climate change. Second, the pesticide impairs their ability to fly and navigate, which hinders foraging and limits food stores. Exposure to systemic pesticides may also damage fertility, as was found in male honey bees.

HABITAT LOSS AND DEGRADATION

As bees struggle to survive, one of their biggest challenges is finding hospitable habitats. Wildflower meadows have been paved over and replaced with housing estates, or ploughed up and polluted by intensive agriculture. And the altered landscapes are full of stressors for bees. Not only have they lost native flora and nesting sites, but they must now contend with pesticides and a lack of floral diversity. In short, the food they have access to is poor quality and there's less of it. Additionally, their once expansive habitat is now divided by roads and vast areas of farmland, which limits their ability to find mates and creates genetically isolated pockets of bee species. These barriers harm the genetic diversity of the species.

CLIMATE CHANGE

Bee life cycles and behaviours are closely linked to seasonal temperatures, weather patterns, and bloom periods. Bees normally emerge from hibernation with the growth of spring flowers and the arrival of sunny days, but changes to our climate have disrupted everything. An early spring could mean that certain flowers will bloom and be gone before the bees have emerged to pollinate them. A late spring may result in a lower survival rate for hibernating bees who could run out of body fat while they wait for warmer temperatures. These mismatches could have major consequences for the survival of both species and all the species that depend on them.

As our climate continues to shift, many species are in danger of not adapting quickly enough. For European bumble bees, who tend to thrive in cooler climates, the rising temperatures are particularly challenging. Many species of bumble bee have failed to migrate north, as other bees have done, and are shrinking in population as a result.

BEE FACT

MANY GARDENERS DON'T REALIZE
IT, BUT FLOWERING WEEDS PROVIDE
POLLEN AND NECTAR FOR BEES.
DANDELIONS ARE PARTICULARLY
VALUABLE TO BEES BECAUSE THEY
BLOOM IN EARLY SPRING WHEN
FORAGE IS SCARCE. THE NECTAR AND
POLLEN THAT DANDELIONS PROVIDE
IS A CRITICAL FOOD SOURCE FOR
HUNGRY BEES JUST EMERGING
FROM WINTER HIBERNATION.

SUPPORTING OUR BEES IN 10 EASY STEPS

There are some small steps that everyone can take to help bees. Below is a list of practical changes that you can make at home to help your local bee population thrive.

1 GROW BEE-FRIENDLY FLOWERS

One of the simplest and most rewarding ways to help bees is to plant flowers for them. With just a handful of seeds, you can create your own little bee haven. If you don't have a garden, you can still support bees by using planted pots on your balcony, by planting in the parkway, or by setting up a rooftop garden. Planting for bees, especially in urban areas, helps to offset habitat loss. For first-time gardeners, try cosmos or sunflowers, which are easy to grow and loved by most bees.

2 PLANT TREES

Most people don't know it, but flowering trees are one of the biggest sources of pollen and nectar for bees. A single mature tree provides millions of flowers, and may continue to do so for decades with little to no maintenance, while pollinator gardens must be managed. Not only do trees provide food, they also help to clean the air and mitigate climate change.

3 INCLUDE NATIVE PLANTS

When planting for bees, make sure to include native plants. The popularity of large-scale nurseries that sell the same plants to customers across a wide range of climates has contributed to a loss of diversity in our gardens. Common imported plants, such as the butterfly bush, are taking the place of native plants that are crucial for some bee species. It's important to include plants that are local to your area if you want to support a diversity of bee species in your garden.

4 LESS MULCH

Although mulch is encouraged by many garden experts, it can prevent bees from accessing the soil they nest in. For a truly bee-friendly garden, make sure you include areas of bare, undisturbed dirt. If possible, designate several sunny patches throughout the garden. Most species of ground-nesting bees have a soil type preference; if your garden has different kinds of soil, providing access to each will give you a greater variety of bee species.

5 PRACTISE NO-TILL GARDENING

With so many ground-nesting bee species searching for habitat, adopting no-till gardening is a commonsense way to support them. Tilling the soil between plantings is a laborious but common practice among gardeners, yet many don't realize that it can destroy active nests or unearth hibernating bees who are not yet ready to emerge. Not only that, but it can harm other beneficial insects in your soil and disrupt the healthy ecology that has been developing over the last growing period. So instead of spending hours churning compost and manure into your garden, simply layer it on top. With no-till gardening, you can save the bees and your back.

6 ALLOW FOR A MESSY GARDEN

Most gardeners like to keep things tidy. They pull weeds and remove dead plants, but doing so could rob bees of important foraging and nesting materials. Flowering weeds such as dandelion, for example, provide crucial early pollen and nectar for bees in spring, often before anything else is blooming. And dead plant stalks and rotting wood offer nesting opportunities for some bee species. Consider leaving some weeds (at least in early spring) and some dried stalks in place. Sometimes a deliberately 'messy' area, like a stumpery, can actually add to a garden's beauty while still providing a nesting habitat for bees.

7 DON'T USE PESTICIDES

When pests get out of control in the garden, don't use a pesticide – reach for the fertilizer instead. Healthy soil can boost a plant's defence mechanisms, fortifying it against pests. In some cases, gardeners may be using pesticides without realizing it. Systemic pesticides, in particular, are often marketed as an 'all-in-one' garden care product that prevents pests and disease while also fertilizing. These products have a long-lasting negative impact on pollinators and should never be used.

8 RESEARCH NURSERIES AND SEED COMPANIES

Even if you've never used a pesticide yourself, if you've purchased plants from a commercial nursery you may have unwittingly introduced systemic pesticides into your garden. Many nursery plants come pre-treated with these pesticides and will continue to produce contaminated pollen and nectar for years. It may take some research, but you can find pesticide-free nurseries and seed companies.

BUY FROM SUSTAINABLE FARMS AND BEEKEEPERS

Many commercial agriculture practices are harmful to bees. For that reason, it's important to support farms and beekeepers that are following sustainable, bee-friendly methods. Even if you are in the city, you might be surprised to find an urban farm or local rooftop beekeeping project. Help businesses like these by petitioning local grocery shops to stock their goods, and by purchasing directly from them at farmers' markets or through community-supported agriculture (CSA) programmes.

BECOME A BEE ACTIVIST

The public has a limited understanding of bees and the complex issues that threaten them. Now that you've read this book, you are no longer one of them. Take action beyond the bounds of your backyard and become a bee activist. This might be as simple as giving an educational talk to your local garden club, or as ambitious as petitioning your city to stop using pesticides. Try volunteering with a conservation group or participating in a citizen science project. You can make a difference. See the Additional Resources section on pages 212–213 for more details.

BEE FACT

WHILE FORAGING FOR FLOWERS, BEES RELY ON SIPS OF NECTAR TO KEEP THEM GOING. THE SUGARS IN THE NECTAR GIVE THEM THE ENERGY THEY NEED TO FLY HOME, BUT WHEN FLOWERS ARE SPARSE, THEY CAN EASILY BECOME EXHAUSTED. IT'S NOT UNCOMMON TO FIND COLD, TIRED BEES ON THE GROUND, WHERE THEY WILL LIKELY DIE WITHOUT SOME EXTRA HELP. IF YOU SEE A LETHARGIC BEE, TRY FEEDING HER A DROP OF SUGAR WATER. TO PICK HER UP, PLACE YOUR FINGER OR A TWIG IN FRONT OF HER AND GENTLY

PUSH TOWARDS HER AND LIFT UP
SLOWLY AS SHE GRABS ON. THEN,
YOU CAN PRESENT HER WITH THE
SUGAR WATER AND WATCH HER
LAP IT. AFTER SHE HAS HAD HER
FILL, SHE MAY REST A FEW MINUTES
BEFORE TAKING FLIGHT. IF SHE SEEMS
UNABLE TO FLY, TRY PLACING HER
IN THE SUN; IT MAY BE THAT HER
FLIGHT MUSCLES ARE CHILLED.

PROVIDING A HOME FOR BEES

Many people endeavour to become beekeepers because they wish to provide a home for bees, but keeping honey bees is not the only option. In fact, establishing nesting sites for other bee species is simpler and probably better suited to most gardeners.

ESTABLISHING A BEE GARDEN

The first step in creating a home for bees in your garden is to establish a food source. If you don't provide flowers, it's unlikely that bees will nest in your garden. Select a wide variety of plants (at least 20 species if space allows) and include natives. When designing your garden layout, you can accommodate bees best by creating patches of each flower type. Most bees only visit one or two flower types per foraging trip and are attracted to large groupings of the same flowers. When selecting which flowers to plant, consider their bloom times. An ideal bee garden always has something in flower, with groupings of flowers that bloom in early spring, late spring, summer, and autumn. Once perennials are in bloom, you should remove dead flowers to stimulate the plant to make new ones.

BEE HOUSES

Now that your garden or balcony is blooming and buzzing with bees, it's time to set up a bee house! These nest-box-like structures are easy to build, require little maintenance, and are unlikely to result in stinging incidents – even in small spaces. They provide nesting opportunities for solitary species such as mason bees and leafcutter bees who are extremely docile and do not live in large colonies like honey bees. Although you can purchase ready-made bee houses, it's recommended that you build your own. Many pre-made bee houses are poorly designed or made with treated wood and will not make suitable homes for bees.

When constructing your bee house, make sure to source wood that has not been treated with chemicals. Start by building the frame. This should look like an open-faced birdhouse: a box with a top, bottom, sides, and a back, plus a peaked roof to protect it from rain. The depth of the frame should be at least 20 cm (8 in), but the other dimensions need not be precise. However, it's best to build a modestly sized bee house. Large houses may look impressive, but they create unnatural nesting conditions for species who don't typically live in close proximity to other bees, and may spread disease. You are better off creating several small bee houses and placing them in different locations around your garden.

Once you've made the frame, it's time to select nesting materials. Bamboo stems or wooden blocks with holes drilled into them (at least 7.5 cm/3 in deep) are popular choices, but you can use any kind of dried, hollow plant stalk. It's a good idea to provide a variety of tubular nesting opportunities with different sized cavities, ranging from 1.5–10 mm ($\frac{1}{16}$–$\frac{3}{8}$ in). The nesting tunnels must be smooth inside, as bees will not nest in rough, splintery cavities. After you've finished preparing your materials, arrange them so that they are tightly packed in your frame.

Hang your finished bee house about a metre off the ground in a sheltered, secure place where it will not be rattled by wind or splashed by rain. Make sure the face of the bee house is not obscured by vegetation or blasted by afternoon sun. As bees begin to use the bee house, you'll notice plugs of mud or leaf matter in occupied tunnels. Inside, the bee larvae are growing.

During winter months, when bees are hibernating, you should remove bee houses from the garden and store them in a cool, dry place, such as a shed, to protect them from moisture. Hang them up outside again in the spring when the bees are ready to emerge. Bees that hatch from your bee house are likely to return to the same location and build their own nests. For this reason, nesting blocks and tubes should be replaced every two years to prevent disease and parasite build-up.

LIVING WITH BEES

Wild bees don't always nest in the spaces we provide for them. Bumble bees, for example, are notorious for rejecting human-made nesting sites. So we must also learn to live in harmony with our bee friends – wherever they choose to nest. They are unlikely to do us any harm, especially when left in peace. It's time to adjust our attitudes towards the bees and think of them not as pests, but as blessings. The next time you spot mining bees digging holes in your lawn or carpenter bees drilling shallow tunnels in the roof of your garden shed, don't ring a pest control company, instead take a minute to marvel. The bees have chosen you.

ADDITIONAL RESOURCES

Also from Hilary Kearney

QueenSpotting

Beekeeping Like A Girl blog

Online beekeeping classes, available at
girlnextdoorhoney.com

Beekeeping

Howland Blackiston, *Beekeeping for Dummies*

Michael Bush, *The Practical Beekeeper*

Les Crowder and Heather Harrell, *Top-Bar Beekeeping*

Kim Flottum, *The Backyard Beekeeper*

Dr Malcolm T Sanford and Richard E Bonney, *Storey's Guide to Keeping Honey Bees*

Thomas D Seeley, *Honeybee Democracy*

Mark L Winston, *The Biology of the Honey Bee*

Honey Bee Suite blog by Rusty Burlew

Bee Craft magazine

Bee Culture magazine

Gardening and Native Bees

Kate Frey and Gretchen LeBuhn, *The Bee-Friendly Garden*

Lori Weidenhammer, *Victory Gardens for Bees: A DIY Guide to Saving the Bees*

Joseph S Wilson and Olivia J Messinger Carril, *The Bees in Your Backyard*

The Xerces Society, *100 Plants to Feed the Bees*

Conservation Organizations
Bumblebee Conservation Trust (USA)
The Great Sunflower Project (USA)
The Pollinator Conservation Association (USA)
The Pollinator Partnership (USA)
Sussex Wildlife Trust (UK)
Urban Bees project (UK)
Wildlife Preservation Canada
The Xerces Society (USA)

ACKNOWLEDGMENTS

Hilary Kearney would like to thank:

This book could not have been written without the dedicated work of entomologists and naturalists, who spent countless tedious hours observing and documenting the life of bees. Thank you for your contribution to the human hive mind.

Nor could this book exist without my editor, Caitlin Doyle, who was so generous in the creative process, considering every idea, no matter how close the deadline was.

To Amy Holliday, it's been an honour to co-create this book with you. Your striking art will be the reason people pick it up off the shelf.

To my mom, for her embarrassing pride in me and my work, and to my dad, for admitting you were proud of me that once.

I'd also like to thank my husband, Tim O'Neil, for making me dinner nearly every night while I obsessively read about bees, and my best friend, Marnie Baird, for making me dinner on the nights that Tim did not.

The publisher would like to thank:

Hilary Kearney for her incredible expertise and boundless passion for bees. Thank you for taking on this lovely project and making it your own.

Amy Holliday for her amazing enthusiasm for the project. Thank you for making our little bee book into a work of art.

Jacqui Caulton, as ever, for her extraordinary design eye and for helping to make a seedling of an idea into a reality.

Thank you to the excellent editorial team of Rachel Malig, Carron Brown, and Helena Caldon. And I am grateful to Meredith Clark at Abrams for getting behind this book from the very beginning.

ABOUT THE AUTHOR

Hilary Kearney is the creator of Girl Next Door Honey, a beekeeping business that offers educational opportunities to hundreds of new beekeepers each year. She is the author of *QueenSpotting* (Storey Publishing, 2019) and maintains the blog *Beekeeping Like A Girl*. Her work has been the subject of features in the *Huffington Post*, *Vogue*, *Mother Earth News*, and other outlets. She rescues wild bee colonies and manages around 60 honey bee colonies in her hometown of San Diego, California. When she's not working bees, she's often found toiling in her large back garden with chickens between her feet. Hilary can be found at www.girlnextdoorhoney.com and at GirlNextDoorHoney on Facebook and Instagram. Girl Next Door Honey apiary tours are offered monthly, and beekeeping classes can be streamed any time. Despite the many fascinating and beautiful bees in the world, the honey bee is still Hilary's favourite.

ABOUT THE ILLUSTRATOR

Amy Holliday is a freelance artist and illustrator based in Cumbria, England. In 2011, Amy won the 'Graduate Showcase' Excellence Award from the University of Cumbria. Since then, she has worked on a wide variety of international projects, including children's and adult books, biological illustration, packaging from food to beauty, and much more. She works primarily in graphite and watercolours, before completing the piece digitally. She spends most of her days in her cosy home studio. When Amy is not working on a commission, she is creating her own work inspired by her latest fascination. Her favourite subjects to illustrate are all things related to the natural world. Other than art, Amy is passionate about wildlife conservation and environmentalism. She runs her own Etsy store, where she sells fine-art prints and products featuring her illustrations. You can find her online at www.amyholliday.co.uk. Amy's favourite bee is the orchid bee, for its beautiful colours.

INDEX